ULTIMATE SUPERCARS

# PORSCHE GT3

By Joanne Mattern

Kaleidoscope
Minneapolis, MN

## The Quest for Discovery Never Ends

........................................

This edition first published in 2023 by Kaleidoscope Publishing, Inc.

No part of this publication may be reproduced in whole or in part without written permission of the publisher.

For information regarding permission, write to Kaleidoscope Publishing, Inc.
6012 Blue Circle Drive
Minnetonka, MN 55343

Library of Congress Control Number
2022938006

ISBN
978-1-64519-614-3 (library bound)
978-1-64519-684-6 (ebook)

Text copyright © 2023 by Kaleidoscope Publishing, Inc. All-Star Sports, Bigfoot Books, and associated logos are trademarks and/or registered trademarks of Kaleidoscope Publishing, Inc.

Printed in the United States of America.

Bigfoot lurks within one of the images in this book. It's up to you to find him!

# TABLE OF CONTENTS

**Chapter 1: Beauty and Speed** ................................................. 4

**Chapter 2: A Long History** ................................................. 12

**Chapter 3: A Peek Inside** ................................................. 18

**Chapter 4: On the Track** ................................................. 24

*Beyond the Book*................................................. 28
*Research Ninja*................................................. 29
*Further Resources*................................................. 30
*Glossary* ................................................. 31
*Index* ................................................. 32
*Photo Credits*................................................. 32
*About the Author*................................................. 32

## Chapter 1
## Beauty and Speed

Adeline and her brother, Matthias, jumped out of the car. "Hurry up!" they called to their parents. The family had arrived at a special showroom. The showroom was filled with cars made by Porsche.

As soon as the family walked inside, Adeline knew they were in car heaven. The big room was filled with beautiful cars. Some were race cars. Others could be driven on the street. All of them looked amazing.

"Adeline, look at this one!" Matthias called. She hurried over to see. "This is the best car I've ever seen!" her brother said. Adeline had to agree.

"I see you've found our new GT3," said Anna, one of the Porsche representatives. "Would you like to know more about it?" Adeline and Matthias nodded, yes.

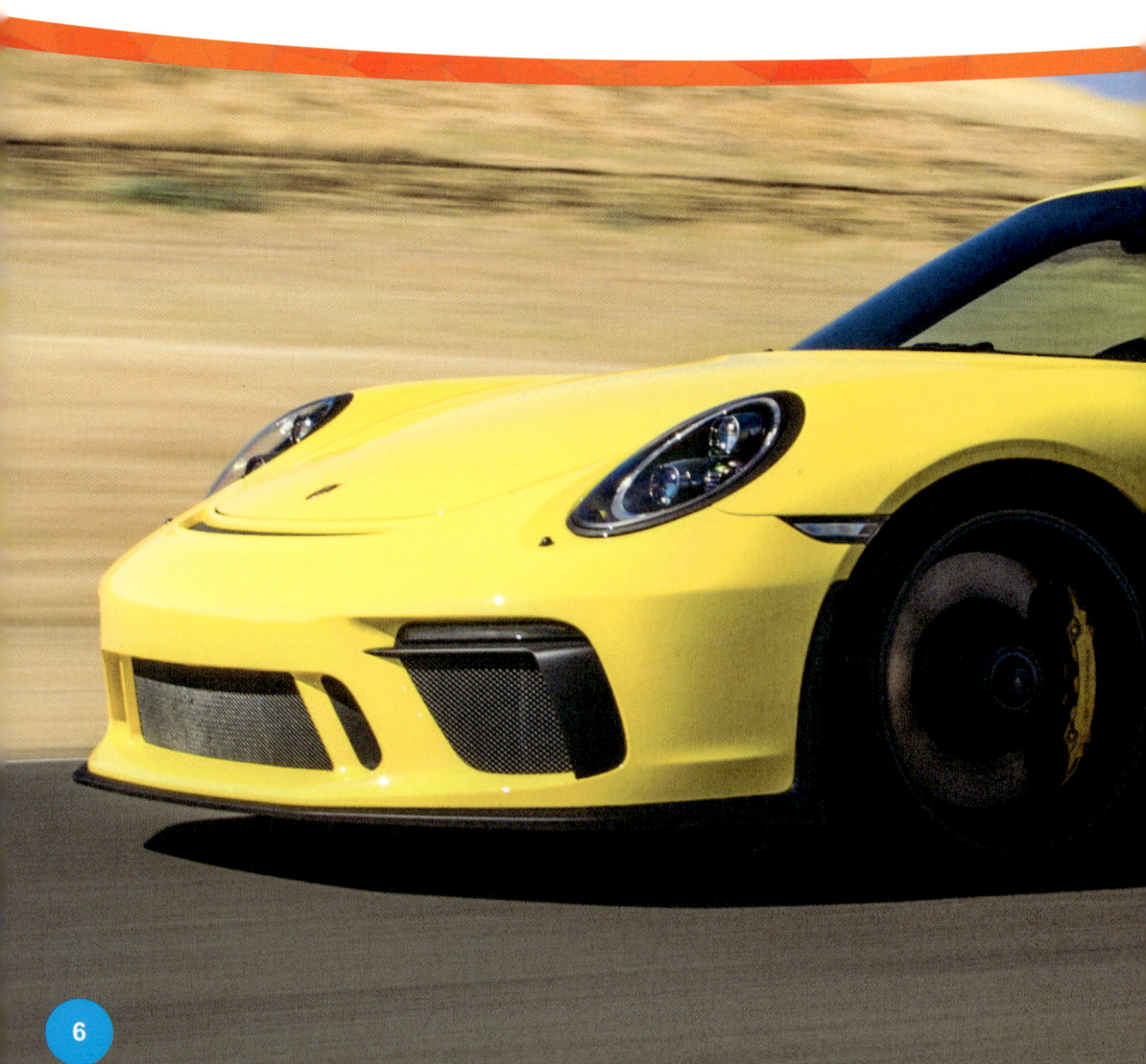

"The fun about this car is you can drive it on the street or on the racetrack," Anna explained. "It is fun to drive and handles wonderfully. But you can also feel its power when you let it loose on the track."

**FUN FACT**
The GT3 can have an automatic transmission with seven speeds, or a manual transmission with six speeds.

"What kind of engine does the GT3 have?" Adeline asked.

"It has a **flat-six** engine. That engine produces up to 502 **horsepower**."

"It looks like a race car," Matthias said. "I love the big wing on the back."

"Yes, that wing makes the car very **aerodynamic**," Anna said. "The car also has tires that grip the road. That makes it easy to drive. You can even go around corners very fast!"

Adeline looked inside the car. "I love the big display," she said. "Is it a touch screen?"

"It is. The driver can control all the car's systems by touching the screen," Anna said. She pointed at the different displays.

"This is a beautiful car," Adeline's father said. "May I ask how much it costs?"

"Prices start at $161,100," Anna said. "That is well worth it to own such a beautiful machine."

# PARTS OF A PORSCHE GT3

rear wing

Sport Cup tires

# KEEPING IT LIGHT

The lighter a car is, the faster it can go. Because the 2022 GT3 has bigger brakes and wheels, it was heavier than the designers wanted it to be. So they made the car lighter in other ways. They used **carbon fiber** for the hood. They also changed the battery to one that is super-light.

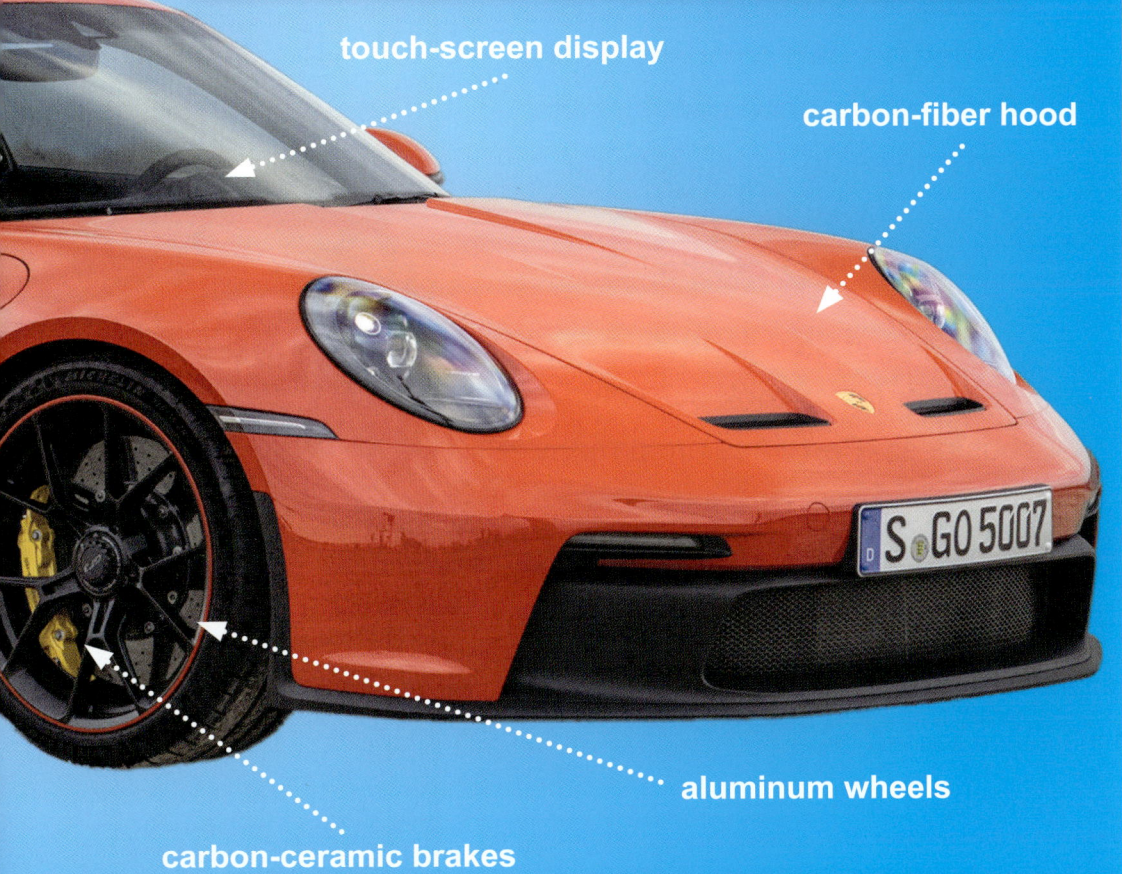

touch-screen display

carbon-fiber hood

aluminum wheels

carbon-ceramic brakes

# Chapter 2
# A Long History

Ferdinand Porsche was born in 1875. He loved cars. Porsche worked for several different car companies. Then, in 1931, he started his own company. At first, Porsche designed cars for other companies. In 1948, Porsche came out with their first car. It was called the Porsche 356.

Porsche died in 1951. His son, Ferry, took over the business. Ferry designed many different cars. The company became known for putting race-car technology into cars ordinary people could drive. That was the idea behind the GT3.

**FUN FACT**
Porsche also designed the Volkswagen Beetle.

The first GT3 came out in 1999. People laughed at the car's funny-looking headlights. They called them "fried eggs." But it did not take long before people came to love the GT3. The car was fast. It was also easy to handle.

Over the years, Porsche added more features to the car. The engine became more powerful. It produced more horsepower. Porsche also made the car lighter. That meant it could go much faster. The 2009 model took just 12 seconds to travel a quarter of a mile.

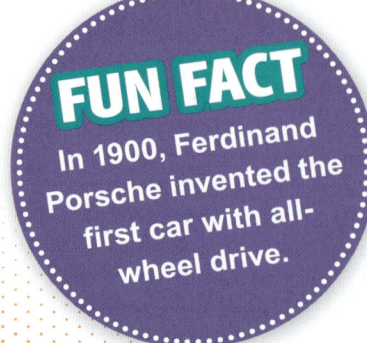

**FUN FACT**

In 1900, Ferdinand Porsche invented the first car with all-wheel drive.

# THE PORSCHE CREST

The Porsche crest appears on every one of the company's cars. Ferry Porsche first sketched the crest on a napkin while eating lunch in 1952. However, the crest is based on an old coat of arms that was once used in part of Germany.

In 2011, Porsche introduced the GT3 RS 4.0. This car became a true collector's item. Porsche only made 600 of the cars. Today, one of these cars costs about $500,000. That's if you can even find one!

In 2014, Porsche stopped offering a manual transmission on the GT3. Instead, the car had a PDK automatic transmission. The PDK made the car even faster. However, some people wanted a manual transmission. In 2016, Porsche made the manual transmission available again. It is still available for the 2022 GT3.

# WHERE IS THE PORSCHE GT3 MADE?

## GERMAN MADE

The Porsche GT3 is made at Porsche's factory in Zuffenhausen, Germany. Porsche's headquarters are also in Zuffenhausen. Their research and development center is in Weissach, Germany.

# Chapter 3
# A Peek Inside

The Porsche GT3 looks amazing on the outside. Anyone sitting inside the car can see it's exciting there as well.

One of the most exciting features of the 2022 GT3 is its display. There is a big analog **tachometer** in the middle of the display. That tachometer is surrounded by many different digital displays. A special GT mode puts all the important information on the digital screens.

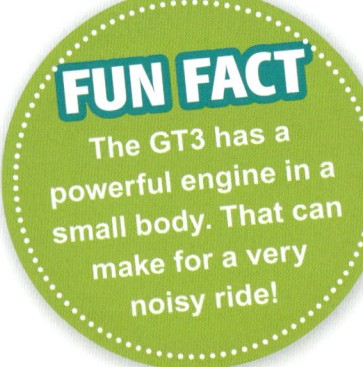

**FUN FACT**
The GT3 has a powerful engine in a small body. That can make for a very noisy ride!

18

In the past, GT3s with an automatic transmission had a stubby gear shift lever. Not in the 2022 model! This model has a gearshift that looks like the handle of a manual transmission gear shift. You can feel the power as you shift into Drive and zoom away.

Of course, the driver and passenger want to be comfortable when they go for a ride. The GT3 has a few different kinds of leather seats. The coolest are the bucket seats. Their light weight and sporty look make them comfortable. They also give the car less weight. That means it can go even faster!

## GOBBLING UP FUEL

Most sports cars get very poor gas mileage. That's because their engines need so much power. The GT3 will need a lot of stops at the gas station. The cars only get about 15 miles to the gallon (6.4 km/liter) in the city. It can get 18 miles per gallon (7.7 km/liter) on the highway.

# THE PORSCHE GT3 IN DETAIL

Height: 4.2 feet (1.3 m)

Width: 6.1 feet (1.9 m)

**LENGTH:** 15 feet (4.6 m)

**WEIGHT:** 3,300 pounds (1,497 kg)

**TOP SPEED:** 199 miles per hour (330 km/h)

**TIME FROM 0 to 60 miles per hour:** 3.2 seconds

If you want to listen to some tunes while you drive, check out the GT3's big touch screen. It is almost 11 inches (27.9 cm) across. Riders can connect to the wireless Apple CarPlay system. The car also can be a WiFi hotspot. Upgrade to a better stereo system for an all-around amazing sound experience.

The Porsche GT3 is a small car. There is some space in the back. However, there is not enough room for passengers to fit in the back seat. Riders will have to sit up front to feel the speed of this beautiful car.

## Chapter 4
# On the Track

**A**nton walked out to the racetrack. He had driven the track at Nürburgring many times. But today was special. Today he would be driving the newest Porsche GT3. Representatives from Porsche were there to watch. They would be filming the ride. They would also time how fast the car would go.

**FUN FACT**
Nürburgring is used for many different races. But ordinary drivers can also pay to drive on it.

Anton climbed into the car. He started the engine and smiled. The car roared with power. It felt like the car couldn't wait to burst from the starting line. Anton shifted into first gear. He liked the crisp feel of the gear shift. It answered his touch so quickly, it was like the car could read his mind.

The car zoomed forward. Anton shifted gears. The engine roared as the **speedometer's** needle moved higher. Anton could feel the tires gripping the road. He felt totally in control even as his speed rose past 100 miles per hour (161 km/h).

The Nürburgring course has it all. It has long straightaways and big curves. The course rises and falls up and down hills. Near the end, the course has a series of sharp turns running downhill. The GT3 handled all of these with ease.

At last, Anton reached the final part of the track. The car whipped around a series of sharp turns. No matter what Anton asked, the car responded.

Finally, Anton reached the finish line. He pressed down on the brakes. The car slowed and stopped. Anton got out. He waited for the Porsche representatives to tell him how he had done.

"Wonderful!" they said. "You reached 199 miles per hour." Anton grinned. He hoped he could take the GT3 for another spin.

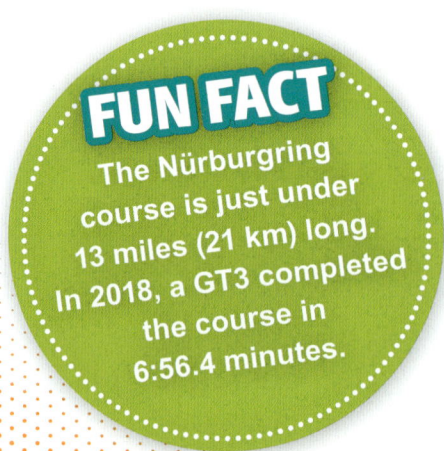

**FUN FACT**

The Nürburgring course is just under 13 miles (21 km) long. In 2018, a GT3 completed the course in 6:56.4 minutes.

## THE PORSCHE EXPERIENCE

Drivers in the United States can drive a Porsche on a track too. Car owners can go to the Porsche Experience Center in Atlanta or Los Angeles. Here, they can get tips on driving the GT3. Drivers can also take the car out on the track for a spin.

# BEYOND
# THE BOOK

After reading the book, it's time to think about what you learned. Try the following exercises to jump-start your ideas.

## RESEARCH

**FIND OUT MORE.** Where would you go to find out more about your favorite cars? Find out what company makes the car and locate its website. What information do the companies provide? What other sources of car information can you find?

## CREATE

**GET ARTISTIC.** Cars start with creative artists and designers. Time for you to take a shot! Get art materials and create a great, new car. Will you make it a sports car? A sedan? A race car? What colors will you paint it? What features can you give it? Let your imagination go for a spin!

## SHARE

**SUM IT UP.** Write one paragraph summarizing the important points from this book. Make sure it's in your own words. Don't just copy what is in the text. Share the paragraph with a classmate. Does your classmate have any comments about the summary? Do they have additional questions about the Porsche GT3?

## GROW

**REAL-LIFE RESEARCH.** What places could you visit to learn more about supercars? What other things could you learn while you were there?

# RESEARCH NINJA

Visit www.ninjaresearcher.com/6143 to learn how to take your research skills and book report writing to the next level!

## RESEARCH — DIGITAL LITERACY TOOLS

**SEARCH LIKE A PRO**
Learn about how to use search engines to find useful websites.

**FACT OR FAKE?**
Discover how you can tell a trusted website from an untrustworthy resource.

**TEXT DETECTIVE**
Explore how to zero in on the information you need most.

**SHOW YOUR WORK**
Research responsibly—learn how to cite sources.

## WRITE

**GET TO THE POINT**
Learn how to express your main ideas.

**PLAN OF ATTACK**
Learn prewriting exercises and create an outline.

**DOWNLOADABLE REPORT FORMS**

# Further Resources

## BOOKS

Cockerham, Paul. *Porsche: The Ultimate Speed Machine*. Broomall, PA: Mason Crest, 2018.

Garstecki, Julia. *Porsche 911 GT3*. Mankato, MN: Black Rabbit Books, 2020.

Mason, Paul. *German Supercars: Porsche, Audi, Mercedes*. New York: PowerKids Press, 2019.

## WEBSITES

**FACTSURFER**

Factsurfer.com gives you a safe, fun way to find more information.

1. Go to www.factsurfer.com.
2. Enter "Porsche GT3" into the search box and click 🔍
3. Select your book cover to see a list of related websites.

# Glossary

**aerodynamic:** an aerodynamic design reduces the drag, or pull, on a car as it moves through the air. The wing on the back of the Porsche GT3 makes it more aerodynamic.

**carbon fiber:** carbon fiber is a very strong, lightweight material. Using carbon fiber to build a car makes it lighter and faster.

**crest:** a crest is a badge or emblem that is a symbol of an organization. Each Porsche car has the company's crest on it.

**flat-six:** a flat-six engine has 6 cylinders. Three cylinders are on each side of the crankshaft. The Porsche GT3 has a flat-six engine.

**horsepower:** horsepower measures the power of the engine. The Porsche GT3 has 502 horsepower.

**speedometer:** a speedometer is a display on the dashboard that tells how fast a vehicle is going. The driver saw on the speedometer that he was going too fast.

**tachometer:** a tachometer is a display on the dashboard that shows how fast the engine is rotating, or spinning. A tachometer displays the rotations per minute, RPMs.

**transmission:** the transmission is the part of the car that moves power from the engine to the wheels. The Porsche GT3 can have either a manual or automatic transmission.

# Index

automatic transmission, 7, 16, 20
brakes, 11, 26
engine, 8, 14, 18, 20, 25,
gear, 20, 25
horsepower, 8, 14,
manual transmission, 16, 20
Nürburgring, 24, 26
screen, 8, 11, 18, 21
tachometer, 18

## PHOTO CREDITS

The images in this book are reproduced through: classic topcar/Shutterstock 16, BoJack/Shutterstock 22, Michele Morrone/Shutterstock 24, David Esser/Shutterstock 27, Mike Mareen/Shutterstock 31; All other images courtesy of Porsche Media/Porsche Newsroom (Hoch Zwei/Juergen Tap 8-9, 18-19, 30; thomas@strogalski.de 21).
**Cover:** Courtesy of Porsche Media/Porsche Newsroom, YIUCHEUNG/Shutterstock (background).

# About the Author

Joanne Mattern has written many nonfiction books for children. Her favorite topics include sports, biographies, animals, and history. Joanne lives in New York State with her family and loves to drive fast cars.